SPACE GUIDES

RING THE
ERSE

GREGO

QED Publishing

QED

D1335680

Sch

80 002 944 572

First published in the UK in 2007 by
QED Publishing
A Quarto Group company
226 City Road
London EC1V 2TT
www.qed-publishing.co.uk

Reprinted in 2008

A catalogue record for this book is available from the British Library.

ISBN 978 1 84835 012 0

Written by Peter Grego
Produced by Calcium
Editor Sarah Eason
Illustrations by Geoff Ward
Picture Researcher Maria Joannou

Publisher Steve Evans
Creative Director Zeta Davies
Senior Editor Hannah Ray

Printed and bound in China

Picture credits
Key: T = top, B = bottom, C = centre, L = left, R = right, FC = front cover, BC = back cover

Alamy Images/Visual Arts Library (London) 10; **Corbis**/Yann Arthus-Bertrand 8–9, /Bettmann 9T, /Jim Craigmyle 29B, /Firefly Productions 28–29, /Simon Marcus 6, /Roger Ressmeyer 10–11, 26–27, 29T, /Howard Sochurek 18BL, /Visuals Unlimited 18BR; **Getty Images**/Hulton Archive 12, /Photodisc FCB, 3, 4–5, 12, 18, 30–31; **Istockphoto**/Dar Yang Yan 19; **NASA** FCT, FCC, 1, 3, 16–17, 18TR, 21T, 24–25, 24, BC, /CXC/M.Markevitch et al 15B, /ESA 23B, /ESA/SOHO 7B, 18LM, /Hubble Heritage 5B, 9B, 22–23, 25, /Hubble Heritage Team/STScI/AURA 27T, /JPL-Caltech/STScI 15T, /JPL-Caltech/ Univ. of Ariz 14–15; **Peter Grego** 13, 17B, 18TL, 23T; **Science Photo Library**/Celestial Image Co 20–21, /David A Hardy 19, /Magrath Photography 12–13, /Emilio Segre/ Visual Archives/ American Institute of Physics 17T; **Nik Szymanek** 27B.

Words in **bold** can be found in the Glossary on pages 30–31.

Contents

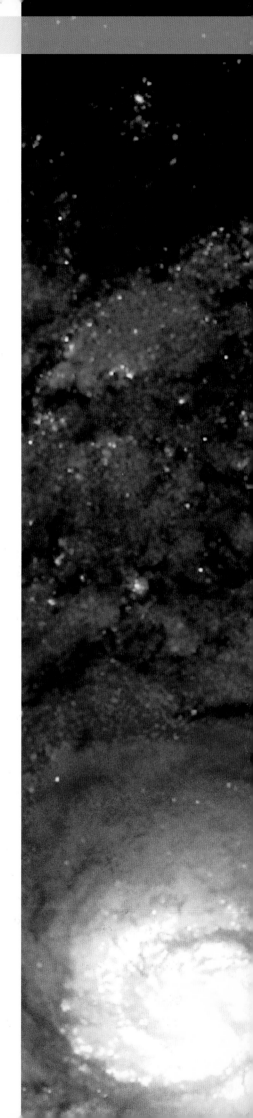

Introducing the Universe

Everything there is makes up the **Universe**. Our own home in the Universe, the **planet** Earth, looks very big to us. But when we look at it from **space**, we realize it is only a ball of rock, one of eight planets circling around a **star** called the Sun.

Our Sun is actually a star. Like all stars, it is an incredibly hot ball of gas. It is so big that more than a million Earths could fit inside it. Now our Earth doesn't seem so big, does it? The Sun and the planets, together with millions of chunks of ice called **comets** and thousands of lumps of rock called **asteroids**, make up our **Solar System**.

⇧ Stars are formed in huge clouds of gas and dust.

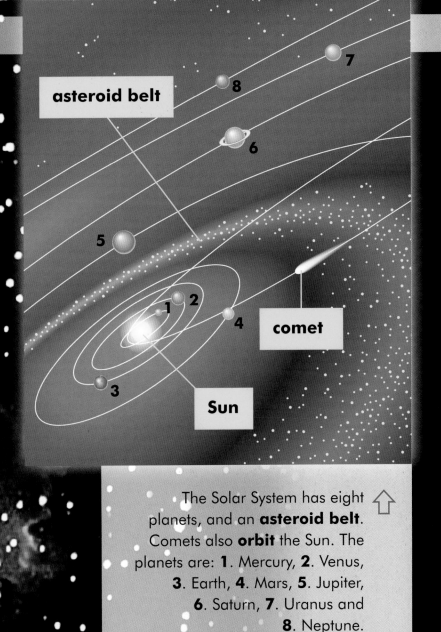

asteroid belt

7

8

6

5

1 2

4 **comet**

3

Sun

The Solar System has eight planets, and an **asteroid belt**. Comets also **orbit** the Sun. The planets are: **1**. Mercury, **2**. Venus, **3**. Earth, **4**. Mars, **5**. Jupiter, **6**. Saturn, **7**. Uranus and **8**. Neptune.

A galaxy of stars

The stars we can see from Earth look small and faint because they are amazingly far away. They lie outside our Solar System and may be bigger or smaller, hotter or cooler, older or younger than our Sun. Many may have their own solar systems, too. All the stars we can see in the night sky belong to a vast **galaxy** called the **Milky Way**. This is a spiral galaxy that is shaped like a fried egg, thick at the centre and thinner on the outside.

A colossal cosmos

The Milky Way is unimaginably big, but it is only a very tiny part of the Universe. There are billions more galaxies in the Universe, some shaped like spirals, others like gigantic footballs.

⬇ The Whirlpool Galaxy, photographed by the Hubble Space **Telescope** – an enormous telescope that orbits Earth.

Stars in their zillions

Amazing

The number of stars in the Universe is estimated to be 70 000 000 000 000 000 000 000. That's 70 sextillion – about the same number as all the grains of sand on all the beaches on Earth!

Universal forces

A few basic forces are in control of the Universe. A force called **gravity** pulls little objects towards bigger objects. For example, when a ripe apple falls off a tree on Earth, gravity pulls it towards the Earth's surface and makes it hit the ground.

Let's play ball with gravity

When you are next playing ball on a warm, sunny day, remember that you owe your fun to the laws of the Universe. Invisible powers are at work! You cannot see gravity, but it is the force that makes your ball curve through the air and fall back to the ground at a certain speed. Earth's gravity also acts on the **Moon**, making it circle around us. The Moon travels once around the Earth each month. Gravity also makes the Earth orbit the Sun in a near-circle. It makes one complete lap in a year. Planet Mars is smaller than the Earth and has just one-third of its gravity. There, because the force of gravity is weaker than on Earth, a ball will fly three times further when it is hit with the same force. Sports rules will need to be different when people play ball on Mars!

⇩ When you throw a ball into the air, gravity pulls it down again. If you could throw the ball hard enough, it would climb so high that it would become a little moon in orbit around the Earth.

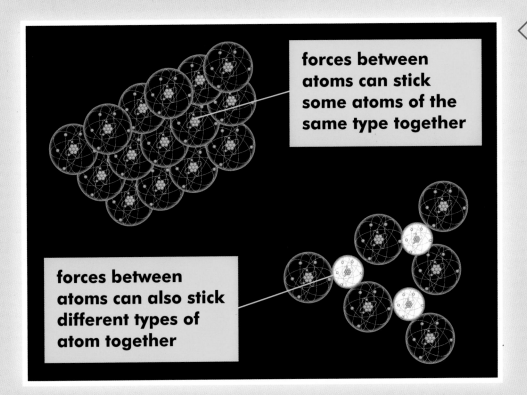

forces between atoms can stick some atoms of the same type together

forces between atoms can also stick different types of atom together

Tiny particles called **atoms** are the smallest pieces of **matter**. Forces inside atoms hold them together. These forces can stick some atoms of the same type together or stick atoms of different types together to form **molecules**.

Atoms make up everything in the Universe. They are like tiny building blocks.

Key Concept

Radiation

All objects in the Universe give out radiation, **which is any form of energy. Light is a type of radiation that we can see, and heat is a form of radiation that we can feel. Other types of radiation need special equipment to be detected. The Sun gives off lots of light and heat, as well as other types of radiation.**

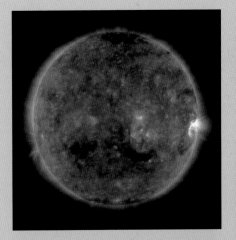
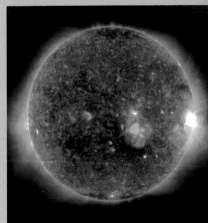
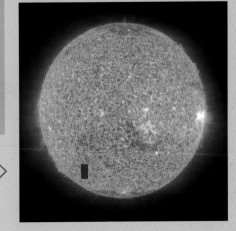
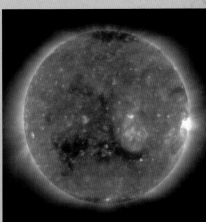

These four pictures of the Sun show different radiation temperatures in the Sun's **atmosphere**. They were taken by an **observatory** in space called SOHO, which studies the Sun.

Cosmic myths and legends

Have you ever looked up at the starry sky and wondered how the Universe began, and where you fit into the big picture? People have been doing this for many thousands of years. Some people have even made up stories about it. One such story was that the Earth was completely flat and surrounded by huge mountains!

The huge Pyramid of the Moon ⇧ was built in Mexico around 2000 years ago. It was created by an ancient civilization that worshipped the sky and Nature.

Strange tales

One ancient Chinese legend tells us that the Universe began as a **cosmic** egg. A god called Pangu grew inside the egg and eventually split it apart. The shell's upper half became the sky, while the lower half became the Earth. When Pangu died, parts of his body fell to the Earth and became the continents.˙

According to an ancient Maori legend from New Zealand, the goddess of the Earth and the god of the sky were once happily cuddled together. Their children grew tired of being squashed in the dark between them, so they struggled and pushed the pair apart. Rain is the tears of the sky god, and mist is the sighing breath of the Earth goddess.

⇧ In ancient Asia, people believed that the Earth was carried on the back of a giant tortoise. This picture shows the tortoise as the foundation of the Universe. The Earth is held up by elephants and the snake represents space.

Unlocking the secrets of the Universe

For hundreds of years, scientists have slowly built up a picture of how the Universe works. Their ideas have replaced ancient myths. For example, they discovered that our planet isn't flat, but instead is a huge globe. People once thought that comets were disaster warnings, but in fact they are chunks of rock and ice that shine brightly when warmed by the Sun.

Mysteries such as this dark keyhole in space ⇧ (a silhouetted **nebula**, see pages 24–25) have been solved by astronomers using powerful telescopes.

An Earth-centred Universe

A philosopher is someone who thinks about problems and tries to solve them. In Ancient Greece, philosophers thought a lot about the Universe. They were on the right track about many things because they were very good at maths, which can be used to find out how the Universe works.

About 2250 years ago, a philosopher called Aristarchus worked out how big the Earth was. He also made a pretty good estimate of the distances from the Earth to the Moon and to the Sun. He said, correctly, that the Earth spins round each day while it orbits the Sun. People did not like this idea. They preferred to believe another philosopher, Aristotle, who said that the Earth was at the centre of the Universe. Aristotle said that the Moon, Sun, planets and stars all move around the Earth at different speeds. People believed this for a very long time, until the telescope was invented in the early 17th century and Aristarchus was proved right.

Aristotle (384–322 BCE) was ▷ one of the greatest of Ancient Greek philosophers, but he was wrong about the Universe.

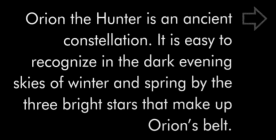

Orion the Hunter is an ancient constellation. It is easy to recognize in the dark evening skies of winter and spring by the three bright stars that make up Orion's belt.

Orion's belt

Constellations

Many ancient philosophers believed that the stars were attached to a giant circle surrounding Earth. People believed they saw **constellations** of gods and monsters, heroes and villains, among the patterns of these stars. To our eyes, the stars look fixed in space. However, they are actually moving, but they appear 'fixed' because they are so far away. Their movements can be picked up by measurements made by telescopes. Incredibly, an ancient Greek would find today's constellations looked almost the same as those seen more than 2000 years ago.

Key Concept

What is a theory?

Scientists use theories to explain the workings of the Universe. But a theory is not just an idea on its own. It fits in with what has been observed, and it can be tested and used to predict how things might happen.

The Sun takes centre stage

About 500 years ago, when most people still believed the Earth was at the centre of the Universe, astronomer Nicolaus Copernicus (1473–1543) wrote a book stating similar ideas to those held by Aristarchus, 2000 years before him. Copernicus said that the Sun was at the centre of the Universe and the Earth was a planet that orbited it. His book upset a lot of people, but it made some people think more scientifically about the Universe, and the modern science of **astronomy** was born.

Biography

Galileo Galilei (1564–1642) and his eye-opening discoveries

More than 50 years after Copernicus's book was published, an Italian scientist called Galileo Galilei turned his small, home-made telescope towards the sky. He saw many things that proved Copernicus was right: the Earth really was a planet orbiting the Sun, just like Mercury, Venus, Mars, Jupiter and Saturn (Uranus and Neptune had yet to be discovered). Proof that the Earth really could not be at the centre of everything came when Galileo saw four tiny moons circling Jupiter every night. This proved that not everything orbited Earth, as Jupiter was clearly orbited by its own moons.

Galileo, a scientific genius, was one ⇨ of the first people to explore the Universe through a telescope.

Galileo made these ink drawings of the Moon from what he observed through his telescope.

Discovering the Milky Way

When Galileo looked at the misty band of the Milky Way through his telescope, he saw that it was made up of countless thousands of stars. Soon, astronomers found that these stars are spread along the flattened disk of our galaxy. We now know that the Milky Way is a spiral shape, and that our Solar System lies far from its centre, in one of its curving spiral arms.

The hazy band of the Milky Way can be seen from dark locations in summer. In the northern **hemisphere** it arches overhead at midnight, making a real treat for those viewing it through binoculars.

Galaxy families

Around 100 years ago, **astronomers** thought that the Universe was only as big as the Milky Way. This idea was shattered when some faint patches of light, dotted about the sky, were shown to be other galaxies. They are as big as our own, but so far away that they appear small and dim through the telescope. Our ideas about the size of the Universe expanded; our own Milky Way now looked rather small.

The spiral galaxy ⇧
Andromeda is the
biggest galaxy in our
'local group'.

Amazing

A near neighbour?

The Andromeda Galaxy, our 'near neighbour', is so far away that its light takes two million years to reach us!

⇦ This is the Large Cloud of Magellan, a galaxy smaller than our own but close to us.

Our galaxy neighbours

Two large patches of light, known as the Clouds of Magellan, were found to be two small galaxies near our own. There are also other galaxies quite close by, at least on a cosmic scale. The Andromeda Galaxy, just visible from a dark location on Earth, is a spiral galaxy like the Milky Way. It is one of the Milky Way's neighbouring galaxies. The Milky Way, the Clouds of Magellan and the Andromeda Galaxy belong to a small group of around 30 galaxies that are connected to each other by gravity.

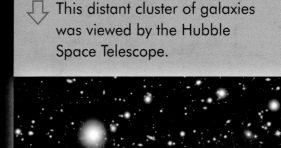⇩ This distant cluster of galaxies was viewed by the Hubble Space Telescope.

Galactic superclusters

Further away, there are many more galaxy families. These families group together to make huge galactic **superclusters**. As we look deeper into space, we peer further back in time. The Hubble Space Telescope has discovered galaxies so far away that their light takes 13 billion years to reach us.

The expanding Universe

The whole Universe is expanding. The further away a galaxy is, the faster it appears to be moving away from us! Imagine that the galaxies are raisins in a cake (the Universe) baking in the oven. As the mixture heats up, it expands and each raisin, or galaxy, moves away from the others. This is how the Universe expands.

Named after astronomer Edwin Hubble, the Hubble Space Telescope has viewed the Universe in great detail since it was launched in 1990.

Edwin Hubble (1889–1953) and the expanding Universe

In the 1920s, an American astronomer called Edwin Hubble noticed that light from far-off galaxies is 'stretched' in an odd way. He realized that the light is being stretched because the galaxies are rushing away from us at enormous speeds.

⬇ Edwin Hubble is shown here with a picture of a galaxy.

The Big Bang

Hubble had discovered that galaxies are rushing away from each other at incredible speeds, but why is this happening? Astronomers now think that it is the result of an unimaginably powerful 'explosion' that happened around 14 billion years ago. This explosion is called the **Big Bang**. Most scientists believe that the Universe was created by the Big Bang. The galaxies are still speeding away as a result of this huge explosion.

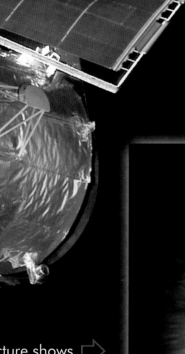

This picture shows ⇨ what the Big Bang may have looked like.

The cosmic calendar

The Universe is unimaginably old. To help us understand the stages in its life so far, let's pretend that its entire 14 billion years are squeezed into a single year. Imagine that the Big Bang occurred and the Universe was created at the very start of 1 January, and the moment you are reading this is midnight on 31 December. Each month of our special cosmic year is just over one billion years long. Each week is about 270 million years long, and each day is 38 million years long.

A long year

You might be surprised to learn that our Sun does not begin to shine until early in August of this year. It is followed a few hours later by the formation of the Earth and the other planets in our Solar System. The first signs of life wait until November to appear, and the dinosaurs arrive on the scene by 24 December.

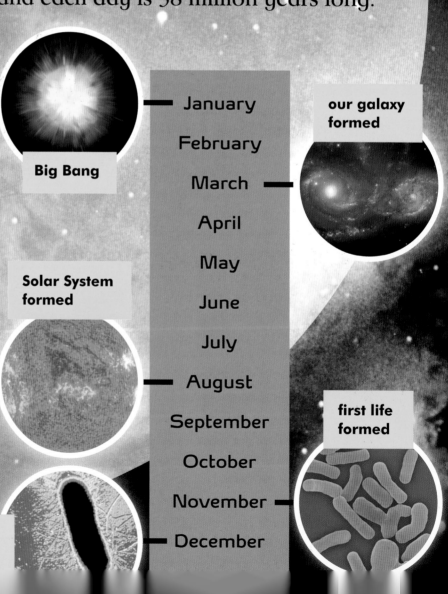

Big Bang

our galaxy formed

Solar System formed

first life formed

complex life formed

January
February
March
April
May
June
July
August
September
October
November
December

December in our cosmic year

1 First complex life	8	15 New life forms flourish	22	29 Dinosaurs wiped out
2	9	16	23	30
3	10	17 First animals with backbone	24 First dinosaurs	31 Mankind
4	11	18 First land plants	25 First mammals	
5	12	19	26	
6	13	20 First animals with four limbs	27 First birds	
7	14	21 First insects	28	

Humans come along at six minutes to midnight o[n] 31 December. The pyramids of Ancient Egypt are built in the last 10 seconds of the year, and just a second before midnight, Christopher Columbus dares to voyage across the Atlantic Ocean. You are reading this book only a very small fraction of a second before midnight, on the ver[y] last day of the cosmic year!

Death of the dinosaurs

Amazing

On 24 December in the cosmic calendar, the dinosaurs took their first footsteps on Earth. Just five days later, a large asteroid smashed into the planet, and most of these fantastic creatures were wiped out in a cosmic instant.

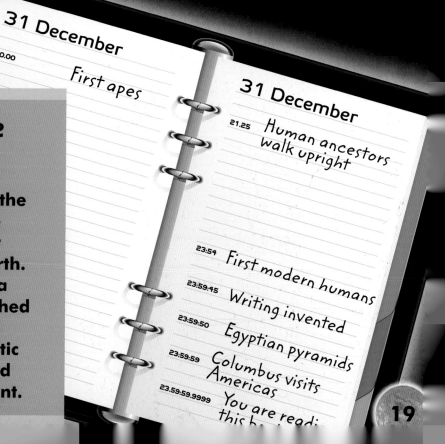

31 December

10.00 First apes

31 December

21.25 Human ancestors walk upright

23:54 First modern humans

23:59:45 Writing invented

23:59:50 Egyptian pyramids

23:59:59 Columbus visits Americas

23.59:59.9999 You are read[ing] this b[ook]

Starry night

Stars are made up mainly of two gases, hydrogen and helium. These gases are so hot that the atoms that make them up move really fast. When fast-moving atoms bump into each other they join together to make a heavier kind of atom, and a burst of energy is produced. This burst of energy, of heat and light, is what makes most stars shine.

You cannot tell how bright a star is just by looking at it, because stars are at different distances from us, and are of different real brightness. A star that looks dim might be much further away than one that looks bright.

This beautiful cluster of young stars is known as 'The Beehive'. ⇨

Star colour

Not all stars are the same colour. Some look blue, others orange. The colour of a star tells us how hot it is. Our Sun is a yellow-white star. It is of medium temperature – its surface is 6000°C. Orange or red stars are cooler than the Sun, while blue stars are very much hotter.

 Stars change colour during their lifetime. When they are young, they are blue; at the end of their life, they become red.

Project

How far the stars?

Hold up a pencil and look at it with just your left eye. Now look at it with just your right eye. The pencil moved! Well, it appeared to move. This effect is called parallax, and astronomers use it to measure distances to the stars. They note the position of a star and six months later, when the Earth is on the other side of its orbit around the Sun, they note its position again. The nearer the star is to us, the more it will appear to move against the distant starry background.

This diagram illustrates the effect of parallax.

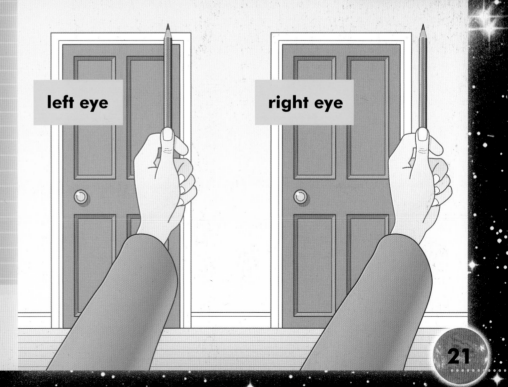

left eye

right eye

Star lives

Small stars burn up their fuel more slowly than bigger stars, and so live longer. As stars run out of fuel, they swell up and their surface cools down. Big, bloated stars like this are known as **red giants**. Eventually they puff away their gases into space. The gas and dust shells around them are called planetary nebulae ('nebulae' means 'clouds', and the shells look like little planets). The stars at the centres of planetary nebulae turn into really **dense** stars about the size of the Earth. These are called **white dwarfs**.

The Helix Nebula is one of the biggest and brightest of all planetary nebulae.

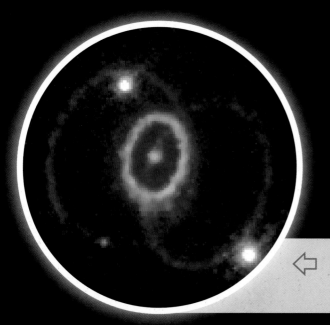

Supernova

Some very big stars last only a few million years before they collapse in the middle and burst apart, in a tremendous explosion known as a **supernova**. Amazingly, most of the matter on Earth was made inside giant stars which exploded a very long time ago.

⇐ This is a supernova explosion. It could make either a pulsar or a black hole.

Pulsars and black holes

A supernova explosion can make one of two kinds of really strange objects – a **pulsar** or a **black hole**. Pulsars are small – about the size of a city – and extremely heavy. They spin up to a thousand times a second! If the star's core is squashed beyond a certain limit, it becomes a black hole. Black holes have so much gravity that everything within a certain distance is sucked into them, even light itself.

Heavy stuff

White dwarfs are so dense that a teaspoonful of their matter would weigh a tonne. That's about the same as a small car. A teaspoonful of pulsar material would weigh a billion tonnes, which is more than all the cars in the world combined!

Amazing

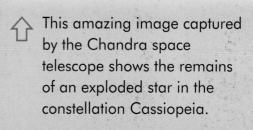

⇧ This amazing image captured by the Chandra space telescope shows the remains of an exploded star in the constellation Cassiopeia.

Gas clouds and fantastic fuzzies

Inside galaxies like our own Milky Way, gigantic clouds of gas and dust stretch for tens, hundreds, even thousands of **light years** between the stars. These clouds block the light from stars behind them, so they appear as silhouettes against the starry background. If you are lucky enough to see the Milky Way at night, you will notice that it is interrupted here and there by these dark clouds.

A star is born

Inside some of the dust and gas clouds, gravity has pulled the material together to make it into dense clumps. Sometimes these become so dense that the enormous pressure and heat ignites stars within them, and the clouds surrounding the stars are lit up like beautiful lanterns. These stars are called nebulae and there are many wonderful examples to see in the night sky.

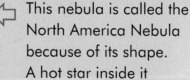
This nebula is called the North America Nebula because of its shape. A hot star inside it makes its gases glow.

A cloud of dust and gas silhouetted against a bright nebula produces the shape of the famous Horsehead Nebula.

The Orion Nebula

The Orion Nebula is one of the loveliest gas clouds in space. It is located just south of the well-known trio of bright stars that make up Orion's belt, and you can see it on clear winter nights and spring evenings as a small misty patch. Using binoculars or a telescope look for wisps of gas and a dark patch of dust known as the shark's mouth. A small group of new stars shining brightly at its centre is called the trapezium.

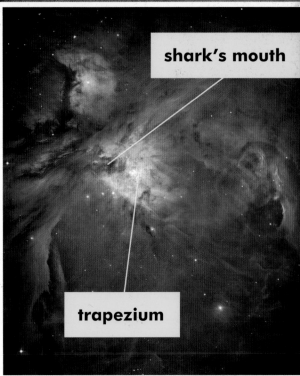

shark's mouth

trapezium

When viewed through an ordinary telescope, nebulae do not look very colourful, but photographs taken through big telescopes bring out their fantastic colours, with blazing blues, rippling reds and a variety of other colours.

Star clusters

As we have seen, stars are usually born in groups inside giant clouds of gas and dust called nebulae. As time goes by, the hot wind streaming from the young stars begins to blow away the nebula around them, until it disappears altogether.

Scattered across the sky are many lovely groups of young stars, some hanging on to traces of the clouds from which they were born. These star groups, known as **open clusters**, slowly get pulled apart as they travel round their galaxy. Our own Sun may once have been part of a star cluster, but it was so long ago that we do not know where its original sister stars are.

Project

Seven Sisters in sight

The most famous example of a young star cluster is the Pleiades (say Ply-a-dees). This cluster is about 100 million years old. It is also known as the Seven Sisters, because keen-sighted people can see seven of these stars with just their naked eye. Look for it on winter nights and spring evenings, by following the line made by Orion's belt, towards the right, by the width of about two outstretched hands.

 The bright Pleiades cluster is a close-knit collection of young, hot stars.

Globular clusters

Globular clusters are far bigger than open clusters. They contain anything from tens of thousands to millions of stars, all held together in a huge ball shape. While open clusters are made up of young stars, globular clusters contain very old red stars, many times older than our own Sun. Around 150 of these impressive clusters surround our galaxy. Other spiral galaxies are surrounded by their own globular clusters.

⇧ Globular clusters are made up of hundreds of thousands of closely packed stars, all held together in a ball shape by gravity.

This is the famous Double Cluster, in the constellation Perseus. It is made up of two open clusters close together. ⇨

Revealing the Universe

About 400 years ago, it was discovered that glass lenses could be used in a tube to make a telescope. This could collect light and give a magnified image of a distant object, such as the Moon. Ever since, astronomers have been making bigger and better lenses, to collect more light and produce more detailed views of the heavens. Telescopes that use lenses to collect light are called **refractors**.

A new kind of telescope

In 1668, a brilliant scientist called Isaac Newton invented a telescope that used a mirror shaped like a shallow bowl to collect and focus light. These telescopes are known as **reflectors**. They can be built much larger than refractors, and they give a clearer image, too. Most of the world's big telescopes are reflectors, using giant mirrors to peer into the distant reaches of the Universe.

 The William Herschel Telescope is a reflector telescope that sits on the top of a mountain peak, on La Palma in the Canary Islands.

Radio telescopes

Astronomers also use radio telescopes. These collect the radio waves from objects in space, in the same way that a satellite dish on your house collects television signals. The signals the telescope collects can be used to make pictures of the objects.

With all these powerful telescopes, astronomers can reveal what is going on in deep space – from the birth of stars, to matter spiralling down the funnels of black holes.

The Very Large Array in New Mexico is one of the world's most powerful radio telescopes. It consists of 27 dishes, each 25m across.

Hundreds of years ago, just a few scientists owned telescopes. Today, many people look at the Universe with their own telescopes.

Telescopes are 'light buckets'

Collecting light is important to astronomers because many objects in the night sky are faint and difficult to see. The bigger a telescope's lens or mirror, the more light it can collect and the brighter and more detailed an object will appear.

Key concept

Glossary

asteroid a lump of rock in space

asteroid belt a band of space between the planets Mars and Jupiter containing thousands of large asteroids

astronomer someone who studies objects in space

astronomy the scientific study of objects in space

atmosphere a mixture of gases surrounding a star or planet

atom the smallest part of any substance

Big Bang the explosion of a single object at the beginning of time, around 14 billion years ago, that created the Universe

black hole a big star that has died and crumpled in on itself, forming an object with so much gravity that nothing – not even light itself – can escape from it

comet a city-sized ball of bits of ice and rock in space. When a comet gets near the Sun, it heats up and produces a long tail of gas and dust

constellation a group of stars within a part of the sky. There are 88 constellations, some of which were formed thousands of years ago

cosmic a word to describe anything in outer space

dense extremely thick and heavy

galaxy a large collection of stars, dust and gas held together by gravity. Galaxies are usually centred around a bulging hub of stars. They can be spiral shaped (like our own Milky Way), football shaped, or have no particular shape at all. Large galaxies, such as the Milky Way, contain hundreds of billions of stars

globular cluster a ball-shaped cluster of very old, red stars containing anything from tens of thousands to millions of stars. Globular clusters are found surrounding galaxies

gravity a force that acts throughout the Universe. The Earth's gravity holds everything to its surface, and the Sun's gravity holds the Earth in its orbit. The bigger and more massive an object, the more gravity it has

hemisphere half the Earth. The top half is called the northern hemisphere, the bottom half is known as the southern hemisphere

light year   the distance travelled by light in one year. Light has a speed of 300 000km per second

matter   all substances and materials. Everything is made of matter

Milky Way   the galaxy which includes our Solar System. Under a clear sky, the Milky Way's more distant stars can be seen forming a beautiful misty band

molecule   a group of atoms. A molecule is the smallest part into which a substance can be divided without changing its chemical nature

Moon   the Earth's only natural satellite. Other satellites are also known as moons (with a small 'm')

nebula (plural: nebulae)   a huge cloud of dust and gas, shining by reflecting light or by being heated from stars within. Stars are born inside nebulae

observatory   a building where space is studied and scientific information about it is collected

open cluster   a group of young stars

orbit   the curved path of a planet or another body round a star, or a moon round a planet

parallax   when the position of an object seems to change when you look at it from a different place

planet   a large, round object orbiting a star

pulsar   a city-sized, very heavy object in space

radiation   all objects in the Universe give out radiation, which is any form of energy, for example, heat or light

red giant   an old star that has swelled up and cooled down

reflector   a telescope that uses a mirror shaped like a shallow bowl to collect and focus light

refractor   a telescope that uses lenses to collect and focus light

Solar System   our cosmic backyard, containing the Sun, the planets and their moons, asteroids and comets

space   everything beyond the Earth

star   a huge ball of burning hot gas

supercluster   a gigantic grouping of galaxy clusters, which are all held together by gravity. Superclusters are the biggest objects in the Universe

supernova   the massive explosion of a large, old star

telescope   an instrument used by astronomers to study objects in space. We can look through simple telescopes to see distant objects in more detail

Universe   everything that exists

white dwarf   a star about the same size as the Earth, made of incredibly dense, tightly packed material

Index